Professional diver's m

PROFESSIONAL DIVER'S MANUAL ON WET-WELDING

D J Keats
Managing Director
Hydromech Technical Services Ltd

ABINGTON PUBLISHING

Woodhead Publishing Ltd in association with The Welding Institute
Cambridge   England

Published by Abington Publishing
Woodhead Publishing Limited
Abington Hall, Abington
Cambridge CB1 6AH, England
www.woodheadpublishing.com

First published 1990

© D J Keats

Conditions of sale
All rights reserved. No part of this publication may be reproduced or transmitted in any form or by any means, electronic or mechanical, including photocopy, recording, or any information storage and retrieval system, without permission in writing from the publisher.

While information is presented in good faith, neither the author, nor the publisher, nor Hydromech Technical Services Ltd warrants the accuracy of information provided or assumes any legal responsibility for it or for any damage which may result from reliance on or use of it or from any negligence of the publisher's, or other persons with respect to it.

British Library Cataloguing in Publication Data

Keats, D J
Professional diver's manual on wet-welding
1.Welding
I. Title
671.S2

ISBN 978-1-85573-006-9

CONTENTS

Preface

Introduction

PREFACE

This training manual contains step by step procedures for performing basic manual metal arc welding operations, as well as information on welding equipment, consumables, and safety.

It is the intention of Hydromech Technical Services to offer training for professional divers in wet-welding and the practice sheets contained in this manual have been designed to enable the diver/welder to progress his new found skill or existing skill further.

In attempting to describe techniques for the welding exercises in detail, it became apparent that some exercises could be accomplished with variations on the technique described. It is acknowledged that welders will adopt variations to suit their degree of skill – this is particularly the case for positional welding where more than one method of electrode weaving may apply.

This manual has been prepared by David John Keats SenAWeldI under a joint agreement with Hydromech Technical Services Ltd, whose policy is to educate and train men in the art of wet-stick welding.

Additional assistance in the form of technical reviews and advice was provided by The Welding Institute, namely, R Spiller, and his colleagues. Our thanks are extended to all who contributed.

INTRODUCTION

Wet-stick welding is one of the methods used in underwater repair and construction. It has been used many times over the years, but unfortunately, not always with great success. One of the reasons for this, is quite simply, lack of understanding and training by the diver/welder and/or contractors.

In this manual you learn in general terms the type of equipment used, factors in planning operations, how best to achieve the desired results, along with the theory and technique to be used when carrying out wet-stick welding.

This manual has been written as a reference book, so you may also like to have it to hand when carrying out your work.

However you wish to use this manual, either as reference only, or as a study manual, the main objective is to enjoy while you learn and do not be afraid to experiment. I would now like to wish you good fortune and hope this manual is of use to you in your career.

CHAPTER 1

INTRODUCTION TO WET-STICK WELDING

Aims of this section
The student should understand and identify:-
the method of wet-stick welding
the advantages and disadvantages
the two basic techniques used
the factors to be considered in planning wet-welding operations

Underwater welding, as the name implies, is welding carried out underwater. The term wet-welding is used to indicate that the welding is performed underwater, directly exposed to the wet environment.

Versatility, speed and low cost makes wet-stick welding highly desirable.

There are practical difficulties however, in welding underwater. Underwater wet-welds are plagued by a rapid quenching effect from the surrounding water, and a susceptibility to hydrogen embrittlement. Both tensile strength and ductility have been found to be reduced compared with similar joints welded in air.

However, having said this, manual metal arc welds made on carbon and C-Mn steels, can be made in water with virtual avoidance of hydrogen cracking with the use of suitable electrodes.

The mechanical properties of the heat affected zone (HAZ) are reasonably high. The limiting factors are on the service viability of a wet-weld, in its strength and toughness and the incidence of weld defects.

Principles of operation
Briefly, the process takes place in the following manner.

The work to be welded is connected to one side of an electric circuit, and a metal electrode to the other side.

These two parts of the circuit are brought together, and then separated slightly. The electric current jumps the gap and causes a sustained spark (arc), which melts the bare metal, forming a weld pool. At the same time, the tip of the electrode melts, and metal droplets are projected into the weld pool. During this operation, the flux covering the electrode melts to provide a shielding gas, which is used to stabilise the arc column and shield the transfer metals. The arc burns in a cavity formed inside the flux covering which is designed to burn slower than the metal barrel of the electrode. A constant arc is maintained even in very poor visibility, the diver exerts a downward pressure on the electrode to keep the flux chipping and burning away to provide constant arc barrel length. A high content of rutile/oxidising/acid/iron powder coatings is often used in underwater welding electrodes. The electrodes themselves are either ferritic or nickel. These two types of electrodes have proved to be the only types that deposit acceptable weld metal.

The major advantage of wet-welding is its simplicity and ability to be used in many different positions on complex joints at the right cost.

9

The incidence of weld defects are dependent primarily upon the consumable type involved. The electrode flux coating has a major effect upon arc stability and weld profile. Arc stability is usually best with rutile or acid rutile coatings which also give welds of generally satisfactory profile and penetration. However, hydrogen cracking is a problem with rutile, basic or cellulosic coatings on ferritic electrodes. This can be minimised with ferritic electrodes with an oxidising coating or by use of nickel based electrodes. With the nickel based electrodes there is a higher susceptibility to solidification cracking, but this can be minimised by additions of manganese.

Therefore, the two ideal wet-stick welding electrodes are nickel based with rutile or acid oxidising coatings, or ferritic with the same flux coating.

Underwater welding techniques

There are two basic techniques used in underwater wet-stick welding:

1. Drag or self-consuming technique;
2. Weave or manipulative technique.

With the drag technique the electrode is dragged across the work and the diver/welder must apply a downward pressure whilst the electrode is being consumed.

With the weave technique, the arc is held as it would be when welding in air, and little or no pressure should be applied. This technique demands a great deal of skill and experience on behalf of the diver/welder.

The drag technique is used in underwater welding more often because in practice a constant contact technique is easier to maintain between the electrode and the work.

This is possible with underwater wet-welding electrodes because the flux covering extends beyond the end of the core wire, and affords an automatic control of the arc length. This also helps prevent the end of the electrode from welding itself, or sticking to the work.

When welding is done using this technique, the weld metal is deposited in a series of beads, or strings, by dragging the electrode along the work. This technique is suited to fillet welding, since underwater it provides a natural groove, so as to help guide the electrode. Fillet beads when laid down in this manner are found to have approximately the same leg length (see Chapter 5 for definition) as the diameter of the electrode.

Factors in planning underwater welding operations

Before any wet-welding operations are started, the job should be inspected to determine whether or not welding can be performed safely and effectively.

Wet-welding is more difficult than dry welding and it requires greater levels of skill.

The main factors to be considered are:

a. Wet-welding can only be successfully carried out by highly skilled and thoroughly trained diver/welders.

b. Specially treated waterproof electrodes must be used and these must have a soft arc behaviour.

c. A platform should be made for the diver to work from, this is of great advantage in strong seas.

d. DC welding generators capable of delivering the required current ratings are necessary.

e. A fully insulated waterproofed electrode holder is required.

f. Correct diving dress – rubber gloves, welding shields, etc.

g. The depth of water to be taken into account.

h. The availability of the necessary tools to complete the job successfully. (Do not try and make do.)

i. The condition and type of material to be welded.

j. Water condition – visibility – sea state – currents, etc.

CHAPTER 2

EQUIPMENT

Aims of this section

The student should understand and identify:-

how the arc is struck
the type of current used
how electrode manipulation can affect arc voltage
the basics of how a welding generator works
the meaning of 'drooping' characteristic
how reverse polarity is obtained
the importance of setting the correct polarity
the difference in UK and USA terminology
the correct procedure for obtaining the correct polarity
the importance of appointing a responsible person for current settings
how water depth can affect current settings
all the personal equipment required
the safety reasons for wearing this equipment
the main safety feature of an electrode holder
how to take proper care of the electrode holder, and understand its safe use
the effect of cable length and voltage drop
the correct techniques used for laying cable out
how earth cable should be attached and/or relocated
a safety switch
how and when a safety switch should be operated

Power supply (general)

The electrical power required for arc welding is far from steady in its demands for both current and voltage. Whenever an arc is struck by causing the electrode to contact the work, a short circuit occurs. The lowered electrical resistance causes a sudden surge of current. A constant current power supply is designed to limit the sudden surges of short circuit currents, thus eliminating the major cause of excessive spatter during welding.

Power sources may supply direct current (DC) or alternating current (AC) to the electrode. However, for underwater wet-welding AC is not used because of electrical safety and the difficulty in maintaining an arc underwater.

A generator supplies only one type of current (DC), but a transformer – rectifier can be switched between AC and DC. The DC welding current arc is more stable than AC.

An important feature of the output is that the current should remain nearly constant during electrode movements, which can vary the arc length. These movements may be accidental or by design to control the weld pool. An increase in arc length will increase the voltage across the arc, but the current should remain near the value selected. Thus the melting rate of the electrode is uniform despite the normal variation of voltage encountered in manual operations. The open circuit voltage (OCV) occurs at the power source terminals, with the machine switched on, but no current flowing.

The welder strikes the electrode on the work to cause a momentary short circuit, when the current is flowing the electrode is very slightly drawn away to establish the arc.

The amperage and voltage for a typical arc length is shown in Fig. 1.

This drooping output is designed to avoid noticeable current variations for unavoidable variations in arc length.

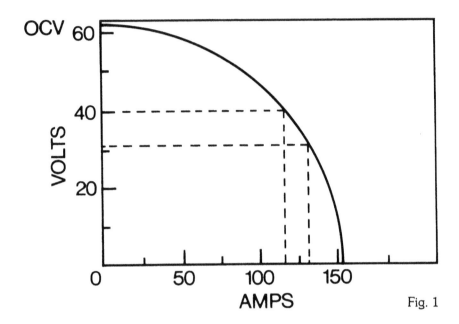

Fig. 1

DC generator

These supply DC with the advantage of more flexible control of the arc characteristics and less current change due to line voltage variation.

The armature in the generator is rotated by a close coupled electric motor, or an internal combustion engine (diesel). Welding current is controlled by varying the current flowing in the generator field windings. Some units are equipped with two types of field control which can be set to provide special arc characteristics for vertical and overhead welding.

The controls are set to reduce the slope of the output to allow a greater than usual change of welding current with arc length and voltage.

On a DC machine the terminals are marked positive (+) and negative (—), except in the case where the polarity can be changed by means of polarity reversing switch. In such cases the terminals are marked "electrode" and "work" with electrode terminal polarity indicated at the polarity switch.

Drooping characteristic

The most widely used type of generator is one where the voltage automatically drops to the arc voltage when the arc is struck and as current commences to flow. This type is said to have a drooping characteristic and methods of obtaining such an effect vary according to the practice of the manufacturer. (See Fig. 1)

Polarity

Underwater welding is done with negative polarity with DC power supplies, Fig. 2. When DC is used with positive polarity, electrolysis will take place and cause rapid deterioration of any metallic components in the electrode holder.

It is, therefore, important to have the correct polarity. Negative polarity is obtained by connecting the negative (—) terminal of the welding machine to the electrode holder, and the positive (+) terminal to the earth clamp. (The earth clamp should be attached to a clean area on the work that will provide a good electrical connection.)

This is not to be confused with the American terminology, which uses different terminology i.e.

Straight polarity is negative electrode
Reverse polarity is positive electrode

If the markings on the generator are not legible, or if there is any doubt, polarity may be determined by the following tests.

a. With the generator dead, connect the earth and welding leads to the terminals.

b. Attach a small metal plate to the earth cable.

c. Insert an electrode into the holder or torch.

d. Immerse the plate and the tip of the electrode in a container of salt water – keeping them about 51mm (2in) apart.

WARNING: ENSURE THAT THE OPERATOR IS PROPERLY INSULATED FROM THE CURRENT.

e. Switch the current on. A heavy stream of bubbles rising from the tip of the electrode indicates negative polarity. If this does not occur, reverse the cable leads and repeat the test.

f. Once negative polarity is established, reversing the cable leads will result in positive polarity.

NOTE: AFTER CORRECT POLARITY HAS BEEN ESTABLISHED, BE SURE NOT TO CONFUSE WHICH CABLE GOES TO THE TORCH AND WHICH CABLE GOES TO THE EARTH CLAMP.

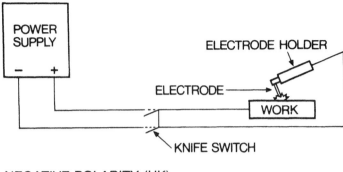

NEGATIVE POLARITY (UK)   Fig. 2

Setting the current

Since the diver/welder will be unable to adjust the welding machine during welding, it is important to determine the welding current setting in advance. This gives approximate current values, which will then only need fine adjustment to give the correct setting.

Up to 20% higher arc currents are needed, due to heat loss from the water. Pressure also affects the size of the gas bubble envelope and it becomes more difficult the deeper you go.

It is also worth remembering at this stage to appoint a specific person who can make the necessary adjustments while you are welding. So make sure he is aware of the welding machine controls and fully understands their function, as this will be of great assistance to you during your welding operations.

Current ratings

Recommended current ranges for underwater welding (guide only)

Electrode size, mm	Position	Current, A
3.25	HV	145-180
3.25	VD	145-180
3.25	OH	140-170
4.00	HV	170-210
4.00	VD	170-210
4.00	OH	170-190

Personal equipment

1. Welding shield.
2. Rubber gloves.
3. Diving dress.
4. Fully insulated electrode holder.
5. Wire brush.
6. Chipping hammer.
7. Scraper.
8. Grinder.
9. Amperage meter (top side).

1. **Welding shield**
   The diver should wear a welding shield fitted with welding lenses appropriate for the water conditions at the site.

2. **Rubber gloves**
   Use of dry rubber or rubberised canvas gloves is **mandatory.**

3. **Diving dress**
   The diver must be fully clothed in diving dress which fully insulates him from all electrical circuits. The suit should be in good condition and free of tears. The exhaust valve button (internal) should be insulated with rubber tape or other suitable means. If hot water suits are to be worn then the diver should wear a rubberised undersuit.

4. **Electrode holder**
   The use of a fully insulated electrode holder designed for underwater use is **mandatory.**

5-8. **Accessories**
   The diver should be equipped with all the necessary tools which can either be pneumatically/hydraulically operated or manually operated. If pneumatic/hydraulic, only approved equipment should be used, and operating instructions from the manufacturers should be carefully followed.

9. **Amperage meter**
   The use of an amperage meter should be made available as the current meters on the welding set may be inaccurate.

   The use of an additional current/amp meter is recommended so that a more accurate picture of the true current/amps may be recorded.

Electrode holder

Durability with increasing depth and insulation for diver protection are the two main features for underwater electrode holders, plus an easy release twist for electrode removal.

Only use electrode holders which have been specifically designed for underwater use and be sure that all current carrying parts are fully insulated with non-conducting material. Springless type electrode holders are recommended whenever possible. Remember, be sure to inspect the electrode holder for worn or damaged parts. NEVER USE DAMAGED ELECTRODE HOLDERS.

Only change the electrode when it is COLD, and never hold the electrode holder so that it points towards you – this is about as dangerous as pointing a loaded gun at yourself.

Special care must be taken not to touch your diving gear with the electrode or any uninsulated parts.

After each day's use, the torch or electrode holder should be thoroughly rinsed in fresh water and dried. This will help to maintain proper operating efficiency.

Power cables

Welding cables – The welding current is conducted from the power supply to the work by multi-strand, insulated flexible copper cable. A return cable (earth) is provided to complete the circuit between the work and power source.

The minimum recommended size is 50mm, because for work that is a considerable distance from the power source, the voltage drop is less because of its lower electrical resistance.

For underwater welding a so-called "whip-lead" of 25mm may be attached to the electrode holder. This will enable the diver to manoeuvre the electrode holder more easily as he welds.

Voltage drops for cables are shown in Fig. 3.

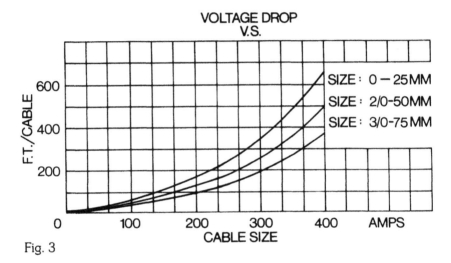

Fig. 3

There will be a voltage drop on a long cable and the output of the welding supply must be adjusted to compensate for these drops. Dirty or loose connections will increase this voltage drop further. To protect personnel from shock hazards and retard corrosion of the connection due to electrolysis, all connections should be fully insulated from the water by several wrappings of rubber or plastic tape. It is also recommended that the supply and earth cables should be as short as possible to minimise voltage drop. The earth cable should be attached to the work as close as possible to the weld joint. As welding progresses the clamp should be re-located to maintain close proximity to the weld joint. It may even prove necessary to use a split earth so as to avoid any magnetic problems which can affect the arc stability.

1. Only use approved cables/connections which are completely insulated and flexible.
2. Inspect all cables for damage before welding.
3. All parts of submerged cables must be fully insulated, and all connections must be made tight.

4. Connect and arrange earth cables so that your body is never between the electrode and the ground side of the welding circuit.
5. Keep welding cables in good condition to avoid unnecessary hazards.
6. Long cables on deck must be laid to protect them from damage and to prevent an unnecessary hazard to personnel. All cables should be laid out in a "kettle element" manner, so as to avoid cable overlap, and this has proved to be useful in helping to avoid magnetic build-up.

Safety switch

The use of a positive acting disconnecting safety switch is mandatory. Knife switches are the only safety switches approved for use.

Greater safety is attained by using a double pole single throw knife switch protecting the welding lead and earth lead sides of the cables (see Fig. 4).

IT IS IMPORTANT THAT THE OPERATOR DOES NOT OPERATE THE SWITCH OR OPEN OR CLOSE THE CIRCUIT UNLESS SPECIFICALLY DIRECTED BY THE DIVER, AND WHEN SO DIRECTED HE SHOULD CONFIRM EACH CHANGE TO THE DIVER VIA THE COMMUNICATION SYSTEM.

1. Current should be off at all times except when you are ready to commence.
2. The knife switch should be placed in such a position that the person responsible can operate it easily at all times.

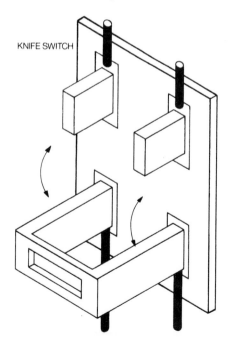

KNIFE SWITCH

Fig. 4

CHAPTER 3

ELECTRODES

Aims of this section

The student should understand and identify:-

the basics of how an electrode is manufactured
how the electrode works
the two types of electrodes used in wet-welding
the basic problems associated with both types
the need for correct handling and care
how to avoid damaging the electrode
the best types of coating for wet-welding electrodes
the function of the coating
why electrodes have a waterproof coating
the requirements of a waterproof coating

Electrodes general

The covered electrode consists of a rod with a concentric coating of flux, the composition of which determines the nature of the electrode. Electrode size is a measure of the rod diameter. The size of the electrode has a major effect on the size of the weld bead, shape and cooling rate. Large electrodes may cause excessive melting of the base metal. The metal transfers from the electrode in small droplets, occasionally a large droplet forms and this short circuits the arc.

The two types of wet-stick electrodes which are dealt with in this manual are:–

i) ferritic
ii) nickel based

The rod is manufactured by cutting wire to length and straightening. The flux is mixed from a variety of minerals and chemicals and then extruded over the rod to form a concentric coating.

The electrode provides filler metal for the weld and conducts electric current to the job. The passage of the current between the electrode and the work piece results in an arc hot enough to melt the electrode and the surface of the work piece (see Fig. 5).

Arc stability is consistently best with rutile/acid/oxidising flux coating electrodes. These also give welds of satisfactory profile and penetration. Basic electrodes give poor arc characteristics and grossly irregular bead geometry, other coating types tend to be between these extremes. Lowest total hydrogen levels are found with ferritic type consumables having an oxidising coating, highest values are with nickel based electrodes.

High nickel wet welds have more porosity that dry welds, but it is evenly dispersed. The nickel electrode gives welds that have good Charpy notch impact strength, and these electrodes have good underwater running characteristics. Both ferritic and nickel electrodes used for underwater wet welds are known to be able to avoid hydrogen cracking in the lower CE steels. (CE = Carbon Equivalent)

Temper bead and high heat input techniques have been evaluated and found to give only limited control of HAZ hardness. In view of the very high cooling rates experienced in wet welds, fully hardened HAZ structures must be anticipated for a wide range of materials.

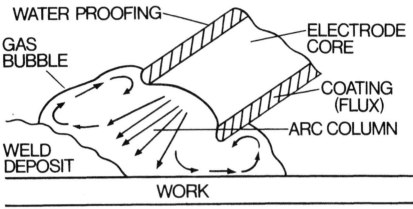

Fig. 5

Electrode identification

Underwater wet-stick welding electrodes are not covered by any BS identification numbers. This is usually left to the manufacturers, so care must be taken to avoid incorrect electrode selection. It is therefore your responsibility to know whether you are to use a ferritic or nickel based electrode. This information will be provided in a welding procedure. You will use the same electrode on the job, as you used on your tests. It is always useful to know what type of electrode you are to use, because this may help you recognise any behavioural differences between the electrodes. You may find one type easier than another to use.

Care of electrodes

It is important always to follow any manufacturer's recommendations for proper care and handling. Remember you are the diver/welder, so care of the electrodes is your responsibility.

To avoid damage, electrodes should be stored in sealed packages in a dry, well ventilated store. Where electrodes cannot be stored in ideal conditions, a moisture absorbent material such as silica-gel should be placed inside each electrode container.

If any electrodes appear to be damaged then they must not be used, always pay special attention to the waterproof coatings for any signs of damage.

Electrodes should always be kept clean, dry and, whenever possible, should be sent down to you as and when you need them. In this way any moisture pick up by the electrodes can be kept to a minimum.

Flux coatings

The best flux coatings on electrodes have been found to be rutile, acid and oxidising types.

Heavy coated rutile and iron powder electrodes give a soft arc and a good contact welding behaviour, but these can give an elongated arc which gives high volts – low amps and so can extinguish.

Development is needed on ferritic electrodes with oxidising coatings and nickel based from the point of view of improving mechanical properties.

The flux coating serves a number of purposes:–

i) to promote easier starting and maintenance of the arc;

ii) when a low thermal conductivity flux is used the flux covering melts slower which forms a protective sleeve over the core wire;

iii) to form and maintain a protective shield around the arc;

iv) heavy slag forming flux coating is desired;

v) it must liberate gases to form the bubble wherein the arc is maintained.

Slag entrapment during welding and slag removal after welding introduce a selection criteria for flux as well:–

i) smaller specific gravity than weld metal;

ii) lower melting temperature than weld metal;

iii) different coefficient of expansion rate than the base metal and weld metal.

The heavier flux coating also helps reduce undercutting due to prolonged heat effect and heat absorption which in turn reduces the cooling rate.

During welding, the flux coating on the electrode melts and deposits slag on top of the weld metal. Flux transfer plays an important part in stabilising the arc by providing an ionised gas stream to help carry the current. It also protects the very reactive metal droplets during the flight, and controls eventual composition of these droplets by necessary alloy additions, scavenges oxides and impurities from the metal, and influences the contour. Finally, the greater the heat absorption of the coating, the deeper the weld penetration, since more time is available for the parent metal to melt before the electrode metal is thrown into the molten pool.

Several reasons can be put forward to explain why the oxidising/iron oxide electrodes prove more suitable for underwater wet welding:–

i) Firstly, the weld metal deposits have a high iron oxide inclusion content and these inclusions by acting as sinks for hydrogen are believed to reduce the diffusible hydrogen content.

ii) Secondly, all weld metal analysis shows that the deposits are virtually pure iron, free from all but traces of the common alloying additions, thus resulting in a lower yield strength, which in turn reduces the levels of applied strain on the HAZ and thus reduces the risk of cracking.

Waterproofing

Without waterproofing the electrode, the coating would deteriorate on immersion into the water, and moisture would be absorbed into the flux. This would cause the flux to crumble and fall off, thus making the electrode totally useless.

The tips of the electrode are ground to a point and covered with the waterproofing agent. This can be epoxy, sealac, celluloid in acetone, waxing agent and other commercial paints, etc.

To start the arc requires a scratch method so as to remove this coating from the tip.

A waterproof coating should be non-conducting and non-hygroscopic to avoid current leakage and moisture pick up, and should burn or fry out easily.

MMA electrode constituents

Constituent	Primary function	Secondary function
Iron oxide	Slag former	Arc stabiliser
Titanium oxide	Slag former	Arc stabiliser
Magnesium oxide	Fluxing agent	
Calcium fluoride	Slag former	Fluxing agent
Potassium silicate	Arc stabiliser	Binder
Other silicates	Slag formers and binder	Fluxing agents
Calcium carbonate	Gas former	Arc stabiliser
Other carbonates	Gas former	
Cellulose	Gas former	
Ferro-manganese	Alloying	Deoxidizer
Ferro-chrome	Alloying	
Ferro-silicon	Deoxidizer	
Binders:–	Used to give the flux covering mechanical strength and help it adhere to the core wire.	
Fluxing agents:–	Used to adjust the surface tension and wetting characteristics.	

Flux coatings
Rutile
Covering	– Main constituents
	Rutile (titanium dioxide) – 50% or more
	Cellulose – Up to 5%
	Fluorspar (calcium fluoride)
Slag residue	– Easily detached, except in deep V
Advantages	– Smooth weld metal surface
	Easy H-V fillets
Disadvantages	– Medium/high hydrogen level
	Limited weld metal toughness

Acid
Covering	– Main constituents
	Iron oxides
	Silica
	Ferro-manganese
Slag residue	– Porous honeycomb – "inflated"
	Easily detached
Advantages	– Good penetration
	Good deposition rate
Disadvantages	– Flat position only
	Risk of solidification cracking

Acid-rutile

Covering	– Main constituents
	Iron oxides
	Silica
	Ferro-manganese
	Rutile or other form of titanium dioxide
Slag residue	– Porous honeycomb – "inflated"
	Easily detached
Advantages	– Good penetration
	Good deposition rate
	Avoids slag entrapment on vertical welds
	Makes best use of limited rutile supplies
Disadvantages	– Risk of solidification cracking

Iron powder

(In addition to normal constituents)

Covering	– Additional constituent
	Iron powder
Operating characteristics	– Efficiency more than 100%
Advantages	– Current and deposition rate may be increased.
	Small core wire allows easy arc control
Disadvantages	– Over 130% efficiency only in H-V or flat positions

CHAPTER 4

HEALTH AND SAFETY

Aims of this section

The student should understand and identify:-
the need for safety requirements
all safety requirements before and during welding operations
the treatment for an electric shock
the correct diving dress

General precautions

Serious injury or death may result if adequate precautions are not taken in underwater welding operations. Personnel should be thoroughly familiar with the safety precautions to be applied when using electrical equipment underwater.

As this manual is concerned with training professional divers to become experienced in underwater wet-welding operations, it is assumed that they have already satisfied health and safety requirements in training with regard to safe diving practice and this manual has not gone into any great depth on the subject. So, all concerned should be constantly on their guard for hazards not covered by this manual.

The explosive gases that are produced during underwater welding are generated from the electrode being consumed and the effects of the water combined. These gases are rich in oxygen and hydrogen and will explode if trapped or ignited.

Precautions check list

1. The electrical power source must be insulated.
2. The welding machine frame must be earthed.
3. All electrical connections must be clean and securely made.
4. Use only electrode holders specifically made for underwater use.
5. Electrode holder and cable joints should be properly insulated.
6. Only change or tighten the electrode holder when no current is in the circuit.
7. Never point the electrode holder towards yourself.
8. Do not let the electrode touch any metallic parts of your diving gear.
9. The current should be off at all times except when actually welding.
10. Always wear rubber gloves to insulate your hands.
11. Always use the appropriate grade of welding filter to protect your eyes.
12. After each dive always inspect your diving equipment for any effects of electrolysis.
13. Be sure that the diver is properly insulated from the current, and all proper safety precautions are strictly obeyed.

Electric shock treatment

General points

Do not touch a shock victim until he has been separated from the current, or the mains supply has been turned off. If you do touch him you may receive a shock. If you cannot turn the current off, use a dry implement made of non-conductive material to separate him from the live apparatus. **ACT QUICKLY.**

When the victim is free check that he is breathing, he may need artificial respiration at once. Try to ascertain the extent of injury, a severe shock will cause burns and even cuts.

Immediately call for an ambulance/paramedic and stay with the victim until the ambulance/paramedic arrives.

What next?

1. Keep a close eye on the patient, make sure he is warm and comfortable and watch for signs of deterioration.
2. If the victim stops breathing then give him artificial respiration until he starts breathing again.
3. If the victim's heart stops beating, then give him heart massage.
4. If the victim's breathing and heart beat recover, then lay him on his stomach, turn his head to one side and draw up the arm and leg of that side, ie place the victim in the recovery position.

Diving dress

1. It is extremely important that the diver's attire affords the maximum protection against any possible electrical shock, and injuries which may be suffered by the eyes from the welding arc.
2. The diver must be fully clothed in diving dress which fully insulates him from all electrical circuits.
3. Dry suits will give the maximum protection, but wet or hot water suits are suitable, but should always be in good condition and free from holes and tears.
4. The use of dry rubber gloves is **mandatory.**
5. The diver should regularly inspect his diving helmet and other metallic parts of his diving gear for signs of deterioration resulting from electrolysis.
6. The diver should never attempt any underwater welding operations without using a welding shield tinted with the appropriate lens both clear and dark.

IN UNDERWATER WELDING, THE IMPORTANCE OF SAFETY MUST BE CONSTANTLY EMPHASISED AND MUST BE IN THE FOREFRONT OF THE DIVER'S MIND.

CHAPTER 5

WELDING TERMINOLOGY

Aims of this section

The student should understand and identify:-
the terms used to relate to parts of a weld
welding terminology

Welding positions

The four basic positions covered by this manual are:–

For fillet welds:–

i)	Flat	(1F)
ii)	Horizontal-vertical	(2F)
iii)	Vertical-down	(3F)
iv)	Overhead	(4F)

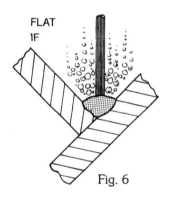

FLAT
1F

Fig. 6

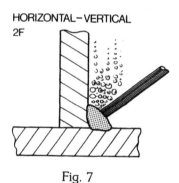

HORIZONTAL–VERTICAL
2F

Fig. 7

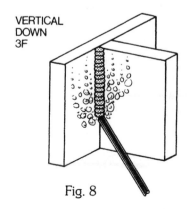

VERTICAL
DOWN
3F

Fig. 8

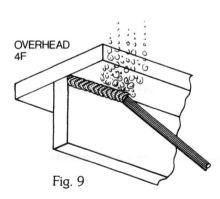

OVERHEAD
4F

Fig. 9

Terms used in welding

When learning to weld it is necessary to become familiar with certain terms that are used.

Term	Definition
Actual throat thickness	The perpendicular distance between two lines each parallel to a line joining the outer toes, one being a tangent at the weld face and the other being through the furthermost point of fusion penetration.
Arc blow	A lengthening or deflection of a DC welding arc caused by the interaction of magnetic fields set up in the work and arc or cables.
Arc voltage	The voltage between electrodes or between an electrode and the work, measured at a point as near as practical to the work.
Covered filler rod	A filler rod having a covering of flux.
Deposited metal	Filler metal after it becomes part of a weld or joint.
Fillet weld	A fusion weld, other than a butt, edge or fusion spot weld, which is approximately triangular in transverse cross-section.
Flux	Material used during welding to clean the surface of the joint chemically, to prevent atmospheric oxidation and to reduce impurities or float them to the surface.
Fusion penetration	In fusion welding the depth to which the parent metal has been fused.
Fusion welding	Welding in which the weld is made between metals in a molten state without the application of pressure.

WELD DETAILS

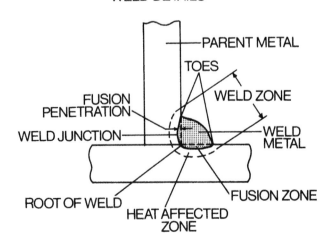

Fig. 10

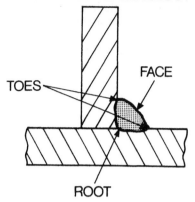

FILLET WELD TERMINOLOGY — Fig. 11

Fusion zone	That part of the parent metal which is melted into the weld metal.
Heat affected zone (HAZ)	The part of the parent metal which is metallurgically affected by the heat of welding.
Leg	The width of a fusion face in a fillet weld.
Open circuit voltage	In a welding plant ready for welding the voltage between two output terminals which are carrying no current.
Residual welding stress	Stress remaining in a metal part or structure as a result of welding.
Root (of weld)	The zone on the side of the first run furthest from the welder.
Run	The metal melted or deposited during one passage of an electrode. Also known as a pass or bead.
Striking voltage	The minimum voltage at which any specified arc may be initiated.
Toe	The boundary between a weld face and the parent metal or between weld faces.
Weld	A union between pieces of metal at faces rendered liquid by heat, or by pressure, or by both.
Weld face	The surface of a weld exposed on the side from which the weld has been made.
Weld junction	The boundary between the fusion zone and the HAZ.
Weld metal	All the metal melted during the making of a weld and retained in the weld.
Weld zone	The zone containing the weld metal and the HAZ.

| Welding procedure | A specified course of action followed in welding including a list of materials and, where necessary, tools to be used. |
| Welding sequence | The order and direction in which joints, welds or runs are made. |

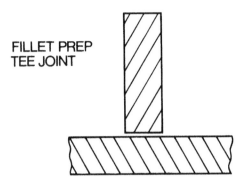

FILLET PREP
TEE JOINT

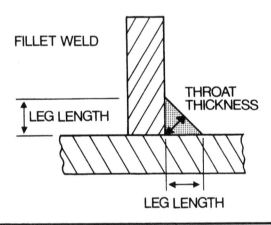

FILLET WELD

LEG LENGTH

THROAT THICKNESS

LEG LENGTH

Fig. 12

CHAPTER 6

METALLURGY

Aims of this section

The student should understand and identify:-

the basic metallurgical effects associated with wet-welding
the effects of hydrogen and solidification cracking
the three structures that dominate in wet-welds, and their effects

Microstructure of wet-welds

Certain metallurgical and physical limitations are inherent in wet underwater welding.

The basic concepts of wet-welding metallurgy include:-

i) The difference in cooling rates between air welds and wet-welds.

ii) The HAZ, grain size and crystal transformation.

iii) Potential underwater defects, quenching induced or hydrogen induced.

When the arc is struck, hydrogen in the arc atmosphere dissolves in the liquid metal of the weld pool. As the weld metal is deposited rapid quenching takes place, leading to swift solidification of the weld metal.

Rapid solidification causes hydrogen to become locked into the lattice of the metal, which then becomes a problem and may cause hydrogen cracking in the hardened HAZ. The rapid quenching effect also results in high hardness and brittleness in the HAZ. Porosity may be formed by this hydrogen and other gases trying to escape. Slag entrapment can also be caused by rapid quenching of the weld.

The water surrounding the weld zone changes the heat transfer pattern during the underwater welding process, and basically three different structures dominate as follows:-

i) **Weld metal** is the zone immediately adjacent to the fusion line, in which metal has been heated almost to its melting point. Because of strong directional heat flow during solidification, dendrites grow more in one direction, therefore long thin grains are a result of this growth pattern. This area is a composition of molten base and electrode metals. On cooling, solidification occurs resulting in formation of tree like columnar grains (dendrites). Occasionally, the heat causes a portion of weld metal to solidify and form a small area of equiaxed grains in the centre of the weld zone, this occurs because the area nucleates itself and solidifies almost instantaneously. These grains grow equally in all directions.

High concentration of impurities in the grain boundaries lowers the melting point. This is extremely detrimental to the welded joint because hot cracking may occur.

ii) **HAZ** is the region most sensitive to fast cooling rates. The maximum temperature in this zone is below the melting temperature, however, this area is severely overheated, allowing a maximum amount of grain growth. There is a grain size gradient across the HAZ, the maximum temperature in the HAZ decreases with increasing distance from the

33

fusion line. During recrystallisation and grain growth the microstructure transformation from austenite to some other crystal structure occurs in those regions where maximum temperature exceeded the low critical temperature.

Once again, the heat flow pattern caused by bubbling on the metal surface influences the cooling. In the region, with maximum temperature above low critical temperature, the temperature is not enough to cause full austenitisation without grain growth, but does cause grain refinement. This subregion in the HAZ is called the annealed region. Reducing heat loss from the welded joint increases the heat affected areas.

It is not the size of the HAZ which makes bad welds, but the grain structure which is responsible for the deterioration of the weld properties. The higher the maximum temperature – the faster the cooling rate. The lower the maximum temperature – the slower the cooling rate. Either stress concentration or hot cracking will make the HAZ weak with regard to mechanical properties. This heat flow due to bubbling will seriously change the grain structure in the HAZ and this is a major reason for failure of underwater welds.

iii) **Base metal (unaffected)** is the area beyond the HAZ which remains unaffected by welding because the temperatures reached are not sufficient to cause any changes. In between the HAZ and this unaffected zone there exists the transient region. Temperatures in this region do not reach the lower critical temperature and therefore no phase changes occur.

Microhardness

Testing the hardness in the different crystal phases, ie pearlite, ferrite, martensite, is also of value.

Hardness values can often be correlated with crystal transformation. Martensite is the hardest crystal structure in steel, the more carbon, the harder the steel.

Pearlite is the equilibrium structure. To determine whether a region will become pearlite or martensite during quenching, the severity of the quench must be known. A more severe quench will result in a larger/deeper hardened zone in a cool plate.

Since a particular value corresponds to a specific microstructural composition, the microstructure is eventually linked through the hardness readings to the cooling rate. The faster the quench, the higher the hardness value.

CHAPTER 7

DEFECTS OF UNDERWATER WELDS

Aims of this section

The student should understand and identify:-
        weld defects
        how these defects can be rectified
        the principles of hydrogen cracking
        the principles of lamellar tearing

Several defects found in underwater welds are also common to air welds, such as arc strike embrittlement, lack of penetration, non-homogeneity of the weld bead metal, gases in the weld metal, slag inclusions and hot and cold cracking.

Defects of welds are of three general classes:–

i) Dimensional

ii) Structural discontinuities

iii) Defective properties

Two basic factors in underwater welding which increase problems with these defects are rapid quenching and hydrogen.

i) Dimensional stresses

Dimensional stresses of high magnitude may result from thermal expansion and contraction and remain in the weldment after the structure has cooled. These stresses cause distortion. Broadly speaking thermal stresses may raise in welded material due to non-uniform temperature distribution. The steep gradient of temperature distribution in underwater welds creates thermal stress build up and warpage.

The turbulence of an underwater welding arc region necessitates stabilisers and gas forming chemicals to stabilise and shield the arc region during welding.

The profile of a finished weld may have considerable effect upon its performance underload. This profile in turn is affected by the viscosity and fluidity of the slag formed during welding. Excessive reinforcement which is usually undesirable tends to stiffen the section and establish stress concentrations.

ii) Structural discontinuties

The rapid cooling effect of water may prevent the escape of gases formed by chemical reactions during welding from the molten puddle. Gas pockets or voids are frequently found in underwater welds. The most frequent reason for porosity is the presence of rust, dirt, oil, paint or other gas producing concentrations in the joint. Thorough surface cleaning must be done before welding underwater.

The flux of the electrode may pick up moisture while welding in water. It goes without saying that water proofed electrodes are recommended. Also there must be sufficient amounts of flux to protect the molten metal from the water, so usually heavier coating type of electrodes are used.

Because of too rapid solidification or too low a temperature during welding, there may not be enough time for slag (oxides) to rise to the surface of the molten metal. When slag becomes entrapped at the fusion boundary, reheat cracking (hot cracking) may become a problem. Obviously, the release of slag from the molten metal will be expedited by any factors that tend to make the metal less viscous or retard its solidification by means of pre-heating, high heat input, slow welding speed or suitable flux protection. Incomplete fusion may be caused by failure to raise the temperature of the base metal to its melting point. To obtain metallurgical continuity of the base and weld metal the surface of the base metal must be brought to fusion temperature. Slag entrapment, subsurface voids, or inproper joint preparation may cause inadequate joint penetration.

If the portion of the base metal closest to the electrode is a considerable distance from the root, conduction of heat may be insufficient to attain fusion at the root. The unfused root permits stress concentrations that could fail without appreciable deformation. The shrinkage stresses and consequent distortion of the parts during welding will cause a crack to initiate at the unfused area.

Undercut occurs when the solidification process takes place too rapidly to allow the weld puddle to recede completely into the toe of the melted puddle region.

iii) Defective properties

Because underwater welding induces an arc atmosphere that is high in water vapour content and in dissociated oxygen and hydrogen, it is thought that hydrogen factors may be especially critical. Hydrogen will not induce cracking unless the region has been hardened and contains residual stresses. Hydrogen is picked up during heating and dissolves in the austenite. As temperature cools down, the hydrogen attempts to diffuse out of weld metal into the air and HAZ. Hydrogen coming out of solution may form or enlarge porosity. Hydrogen that has super saturated the metal may result in cracking by the following mechanisms:-

1. The hydrogen diffuses to areas of stress concentration, such as areas of martensite structure.
2. The stressed area causes a crack initiation after the introduction of hydrogen, then the crack is allowed to propagate.
3. The crack grows in steps.

Hydrogen cracking

Wet-stick welding produces rapid heating and cooling in the HAZ. On steel this causes hardening in this zone. The wet-welding process releases high levels of hydrogen into the arc, and therefore into the molten weld pool. When the weld cools the combination of:−

i) a hardened zone
ii) hydrogen in the weld and HAZ
iii) stress

may cause cracking.

Fig. 13

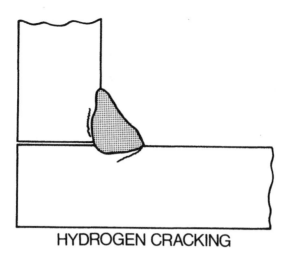

HYDROGEN CRACKING

Typical hydrogen assisted cracking is shown in Fig 13. It may be below the joint surface and needs NDT for its detection.

Although hydrogen assisted cracking usually occurs in the HAZ of the parent metal, it is also known to occur in weld metal.

The risk of hydrogen cracking depends to a large extent on the material composition, since this determines the type of hardened structure formed in the HAZ. Materials of high carbon content are more prone to hydrogen cracking because of the susceptible hardened HAZ.

The cooling rate is also of importance, as fast cooling rates produce an HAZ which is more prone to cracking than a softer HAZ, which is produced by a slow cooling rate. Most of the heat of the cooling weld is extracted by the surrounding water and cold metal.

Lamellar tearing

Lamellar tearing is a type of cracking which occurs in the parent plate near the weld. It occurs when non-metallic inclusions in steel plate create a path of weakness which cracks during the contraction of a cooling weld on the surface of the plate. The microscopic inclusions are rolled into flat shapes during the hot rolling of the plate by the steel maker. Since all these flattened shapes lie parallel to the surface of the plate they will only cause a path of weakness if the weld contraction acts in the thickness direction of the plate. Thus the usual joint type affected is a T butt joint or a T joint with a large fillet weld.

The characteristic zig-zag pattern of the crack seen in cross-section results from the tendency of the crack to follow the flattened inclusions which are at different levels (see Fig. 14).

Fig. 14

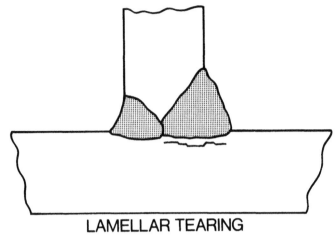

LAMELLAR TEARING

In most circumstances the contraction of the T joint weld occurs without lamellar tearing as the parts of the joint move or distort to relieve the contraction stress. If however the parts are heavily restrained, high thickness direction stresses will develop in the cross plate of the T joint and lamellar tearing may occur if there is a collection of inclusions near the weld fusion line.

Hydrogen in the weld will increase the risk of cracking and precautions against hydrogen cracking will assist in avoiding lamellar tearing.

Technique defects

The defects listed below are generally those associated with faulty technique or poor workmanship:

a) **Undercut (Fig. 15)**

Cause	*Prevention*
Long arc	Keep shorter arc. Watch angle of electrode. Lower o.c.v.
Restricted access to joint preparation	Allow more room in joint for electrode manipulation. Use smaller electrodes.
Incorrect angle of electrode	Correct angle between 30 and 45°

b) **Lack of fusion (Fig. 16)**

Cause	*Prevention*
Incorrect electrode angle	Correct angle between 30 and 45°
Travel speed	If too fast does not allow time for proper fusion
Low current	Increase current (with caution)
Welding over gaps	Better preparation techniques required

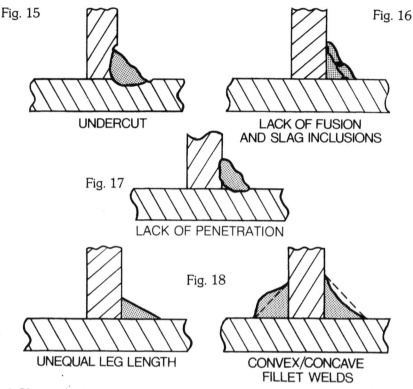

Fig. 15

UNDERCUT

Fig. 16

LACK OF FUSION
AND SLAG INCLUSIONS

Fig. 17

LACK OF PENETRATION

Fig. 18

UNEQUAL LEG LENGTH

CONVEX/CONCAVE
FILLET WELDS

c) Slag inclusions

Cause	*Prevention*
Careless or incorrect cleaning	Extra care to be taken to ensure all slag is removed
Irregular or convex welds	Adjust welding conditions to obtain smoother weld deposit
Lack of penetration with trapped slag beneath weld	Use smaller electrode or increase amperage

d) Incomplete penetration (Fig. 17)

Cause	*Prevention*
Current too low	Increase current (with caution)
Electrode diameter too large	Use smaller electrodes
Incorrect electrode angle	Correct angle between 30 and 45°

e) Incorrect weld profile (Fig. 18)

Cause	*Prevention*
Excessive weld concavity	Improve run placement
Unequal leg lengths	Adjust run placement
Concave fillet weld	Reduce speed, reduce amps

CHAPTER 8

UNDERWATER WELDING PROCESS PARAMETERS

Aims of this section
The student should understand and identify:-
>
> the parameters that control weld quality
> arc behaviour
> factors that influence the final weld microstructure
> travel speed and its effects
> the importance of correct electrode size and type
> the effect of water on weld quality
> the effect of increasing depth on the arc
> the effects of magnetism and how it may be prevented, or
> overcome
> how to demagnetise

In order to control the quality of welds, it is necessary to know the parameters of a specific welding process.

The following have been found to be most relevant:–

i) Characteristics of the arc – current, voltage and polarity.

ii) Base metal, welding speed and lead angle.

iii) Electrode size and type.

iv) Water environment – pressure and water characteristics.

v) Magnetism.

Arc characteristics
The welding arc does not behave underwater as it does in the air and the activity of the gas bubble is particularly important when the arc is struck. The combustion of the electrode and the dissociation of water creates a gas bubble. As the pressure within the bubble increases, it is forced to leave the arc and meet the surrounding water, while another bubble forms to take its place. Therefore if the electrode is too far away from the work, the weld metal will be unable to transfer in an acceptable manner. If the electrode is too close, the bubble will collapse around the weld and destroy the possibility of producing an acceptable weld.

In general, the voltage in underwater welding needs to be higher to strike the arc, than it does to maintain the arc. The amperage grows as the voltage decreases, after the arc has been established. Since voltage drop depends upon the electrical resistance of the arc gap, arc length, and chemical composition surrounding the arc, electrical field strength in the underwater arc increases with increasing current.

i) Arc current decreases with increasing water depth, and therefore voltage increases in order to maintain the same amount of heat input.

ii) Due to the fast cooling rates, more heat dissipates into the surrounding water and more heat input is necessary. Thus higher voltage is required.

iii) The high hydrogen content in the arc zone increases the electrical resistance. Thus higher voltage is required.

41

Welding machines have a specific relationship between current and voltage, with the current being the only independent variable. It is also true to say that in salt water the welding arc will become more stable due to salt ions serving as charge carriers. This is evident in the change of sound of the arc from an erratic effervescence in fresh water to a smooth gurgling in salt water.

Finally, it is mentioned that depending on the quality of equipment used, there can be quite an effect on changing what actually happens to the arc, compared to what is thought to happen. Polarity of the current can be a major factor in controlling the bead appearance and penetration. In general, 80% of the heat goes to the anode, 15% is carried away by gases/flux, and 5% to the cathode. DC reverse polarity is a better setting for underwater welding. It is noted that DC straight polarity is detrimental to the electrode holder, but does result in better welds.

Due to the uncontrollable heat losses from the welding arc, the actual heat that reaches the weld is a percentage of the heat input. The value of this factor in underwater welding is not known precisely, but heat input to the work piece will determine the amount of melting and the penetration of the weld bead and also the distribution of maximum temperature in the vicinity of the molten puddle.

The size of the bubble generated from the arc during underwater welding is directly influenced by the heat output from the arc. Higher heat input is usually required to generate large bubbles and less frequency of bubble departure to shield the weld from the water environment.

A slower welding speed is recommended for underwater welding.

The heat which flows away from the weld zone determines both the time that the region stays at its maximum temperature and the cooling rate from this temperature. Both these characteristics are critical in determining the final weld microstructure.

Travel speed

Welding speed is a function of both the electrode and current setting, forcing the electrode to go faster will usually result in an uneven, sporadic weld with a narrow peaked bead with no improvements in penetration. In "drag" arc welding the welding arc digs out the weld crater at a specific rate. Thus the "drag" technique leads to a natural characteristic welding speed. A slow welding speed is required in underwater welding. But, it must be realised that the size of bead will be determined by the size of electrode used, and the travel speed itself is, to a point, self setting when the diver/welder has learnt to master the skill.

In general, the smaller the lead angle the wider the weld bead and the shallower the penetration.

The rate of travel will depend on:—

> welder skill
> electrode size
> the current
> the run size required

Electrode size and type

With regard to electrode type, we do not have a great deal of choice because as mentioned earlier there are only two types of electrode that can be used successfully underwater in wet-stick. These are:

i) Ferritic

ii) Nickel based

Selecting the right size of electrode is also of importance because we need to control factors such as current/volts, travel speed, lead angle and arc bubble generation, as much as possible. We do not actually have a very wide choice and the range of sizes are between 3.25-5.00. 4.00mm is probably the best choice that the diver/welder can control satisfactorily. Any larger and the current needed is getting excessive and the demands on the diver/welder to make a satisfactory weld are becoming higher. Any smaller, and the amount of transferred metal is unrealistic due to fit-up problems, and manipulation difficulties.

Water environment and its effects

Theoretically speaking an electric arc burning underwater is no different from other types of arcs.

However, underwater conditions, including the static pressure caused by depth, the cooling effect of the water, moving the electrode, and high hydrogen levels, cause a rising voltage/amp characteristic for underwater arcs.

In addition to the hydro environment effect on the arc, there are variables in visibility, water movement, diver stability, etc. Underwater welds exhibit only 80% of the tensile strength and 50% of the ductility of welds in air.

There is not enough time for slag (oxides) to come to the top of the molten weld metal during fast solidification. These entrapped slags cause non-uniform solidification and high residual stresses which cause crack problems. The mobility of the molten bead is restricted by the environment and temperature. The weld bead usually appears to be convex.

The decomposed hydrogen near the arc during underwater welding may become entrapped and cause porosity. This decreases the impact toughness and leads to hydrogen induced cracking.

A contaminated water environment may cause serious porosity problems during underwater welding.

The water environment offers more chance of decomposition of water into hydrogen and oxygen. To improve weld quality we need to avoid rapid cooling, control solidification, control electrode melting and improve diver training.

Magnetism

Electromagnetic forces can build up in a work piece during wet-welding. Eventually they can prevent arc initiation and there may have to be a limit to the welding time on a large job.

When wet-welding offshore the oil rigs themselves are protected against corrosion by sacrificial anodes which are charged electrically and the rig has actually got its own magnetic field. So these must be turned off before any welding can start, but there may still be residual magnetism left which may cause the diver/welder problems.

During welding operations the diver will probably not notice the arc wandering from side to side due to the bubbles coming from the arc, or poor visibility.

The arc moves in this way due to magnetic influences, and it is impossible to overcome the behaviour of the arc. Therefore two essentials are needed:–

i) Measures to help prevent magnetic build-up.

ii) Effective demagnetisation techniques.

To help prevent magnetic build-up is very difficult, however, certain techniques can help:–

a) Use short run placements.

b) Weld towards the earth clamp.

c) Use two earth clamps if needed.

d) Do not build up welding passes too much or build them in one area before another. Spread the welding around all the joints to be welded systematically.

e) Avoid tangling the cables around each other and the job.

f) Use as low a current as possible.

Effective demagnetisation techniques

If a magnetic build-up has occurred and is becoming a problem, then prior to trying to demagnetise, it is worth trying to even the magnetic forces out by wrapping the earth cable around the top of the joint and the welding cable around the bottom (see Fig. 19).

Or, a change in polarity may help, but we do not recommend this due to electrolysis and the fact that straight polarity is not as safe to use.

Having tried all these points without success, either abandon the dive for the day and demagnetise before welding the next day, or demagnetise there and then and continue. There is, of course, specialist equipment available designed to demagnetise, or if any MPI has been taking place, then this equipment can be used to demagnetise by wrapping the cables around the joint and subjecting the component to a continual reversing magnetic field.

There is, however, no guarantee that any of the above steps will prove fully successful, but they should greatly alleviate the problem.

MAGNETIC LINES
OF FORCE

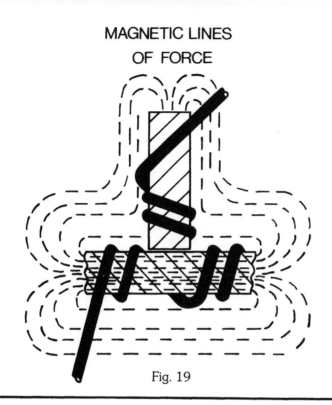

Fig. 19

CHAPTER 9

WELDER QUALIFICATIONS/WELDING PROCEDURES

Aims of this section

The student should understand and identify:-
> the difference between the two tests
> the reasons for such tests

Welder approval qualifications

What is a welder approval test?

This is a test or a series of tests in which an individual welder welds a test piece under specified conditions. This test piece is then examined by non-destructive and/or destructive tests in accordance with a national standard or an internal company standard.

If the weld meets the desired standard then the welder is approved to weld on work of the same type as the test and usually a range of similar welds.

Here is a list of some standards commonly used.

British Standard	BS 4871, 4872
American Standard	API 1104
	ASME 1X
	AWS D3.6 – 83
French standard	Bureau Veritas
Norwegian standard	Det Norske Veritas

At the present time, in the UK at least, the test most likely to be called for would be the AWS or possibly Det Norske Veritas.

As the diver/welder you will be given written instructions which will detail the following information:

a) The parent material
b) Plate, pipe branch connection, etc
c) Welding process
d) Type of consumables
e) Position
f) Material thickness
g) Type of joint
h) Weld dimensions

Other factors may be left to the welder's choice, ie electrode size, run sequence, travel speed, etc.

Note: A session will normally be given to allow the diver/welder to familiarise himself with what is to be expected.

Now let us take a closer look at our written instructions.

a)   Any material being used for any tests must first be identified, this can be done in various ways:

1. by specification
2. by manufacturer's trade name
3. by chemical analysis

Most weldable structural materials are classified in terms of the chemical composition and mechanical properties, example BS 4360 specification for weldable structural steels.

Many welder approval specifications allow for materials to be put into groups because of similarities on a weldability factor.

b)   Most approval standards group metal products into three main groups:

1. sheet material
2. plate
3. pipe

Most standards specify the length, width, thickness and outside diameter of the test piece – (pipe)

c)   For the purposes of wet-stick welding the welding process is manual metal arc (MMA).

d)   Electrodes are usually classified in terms of chemical composition and mechanical properties. BS 639 1976 covers electrodes for manual arc welding of carbon and carbon-manganese steels. However, there is no BS specification for wet-stick welding electrodes, these will usually be sold under a trade name, but will either be ferritic or nickel based.

e)   The position is usually indicated by abbreviations (F) for fillet welds and (G) for butt welds. For fillet welds refer to Chapter 5. Butt welds are not covered in this publication.

f)   Most approval tests will group the thickness of a material so the welder is approved on more than just one thickness.

g)   Although there are a number of different types of joints this manual only deals with fillet welds on T or corner joints and this will be your most common joint type on any tests. Edge preparations for all fillet weld tests will be square edge.

h)   This information refers to the size that the finished weld must be and is measured by its leg length and/or throat thickness.

Most approval tests will allow for the welding of two samples. If the welder for whatever reason feels that his first sample is not satisfactory he will be allowed to have a second attempt. However, it must be pointed out that the first sample will be disregarded and the second sample will be the one that has to be tested.

Welding procedure

A welding procedure differs from a welder approval test in that in the procedure test it is not the welder's skill that is being tested, but whether the particular details that the welding engineer has calculated, will in fact be fit for their purpose. Simply put, it concerns whether it is possible for a sound weld to be produced by the instructions given.

It goes without saying that the welder chosen to attempt the test will be one of a high calibre.

If the welding procedure test is successful then the welder who conducted the test will receive a welder approval qualification, as well as the company receiving their welding procedure certificate.

Let's just confirm what exactly has happened then. The welding engineer got all his information correct, the welder therefore was successful in producing a sound weld, so the company got a welding procedure approval, and the welder got a welder approval qualification.

Welding procedure approval

A welding procedure is therefore a document that gives the precise details of the steps by which the welding of a certain joint is to be carried out. It contains the necessary details, or range of details, for the control of the variables used in producing the welded joint. The procedure thus controls the quality of the weld by controlling each step that is required to be taken. Here is a list of the essential details:—

a) Welding process or processes.

b) Material specification (thickness, OD, ID, etc).

c) Site or shop.

d) Edge preparation.

e) Method of cleaning.

f) Fit-up.

g) Jigging or tacking.

h) Welding position.

i) Type of consumable.

j) Filler material, composition and size.

k) Pre-heat temperature method of control.

l) Travel speed.

m) Arrangements of runs.

n) Welding sequence.

o) Back gouging or not.

p) Post-weld heat treatment.

q) Drying temperature/times for consumables.

r) Any specific features – heat input control, eg run out length (ROL).

CHAPTER 10

NDT AND DESTRUCTIVE TESTING FOR QUALITY CONTROL

Aims of this section

The student should understand and identify:-

the difference between non-destructive testing (NDT) and destructive testing
the defects which can be assessed by visual inspection
the basic principles of magnetic particle inspection (MPI)
the basic principles of ultrasonics
the basic principles of radiography

Tests for weld quality are usually carried out by NDT methods, while tests for mechanical properties are usually determined by destructive tests.

Visual inspection

Visual inspection during and after welding will detect surface defects, while a trained observer may also detect clues to internal defects.

The weld surface contour should have a linear and uniform profile, no excessive undulations, a consistent weld ripple shape, and no excessive overlap.

Visual inspection will find defects such as:-

i)   Blow holes/worm holes which come to the surface.

ii)   Imperfect shape – overlap – unequal leg length – poor restart.

iii)   Surface porosity.

iv)   Excessive weld reinforcement.

v)   Undercut.

vi)   Gross cracking, crater cracking, longitudinal cracking, etc.

vii)   Miscellaneous faults – stray arc flash – spatter, etc.

(See Fig. 20 and 21)

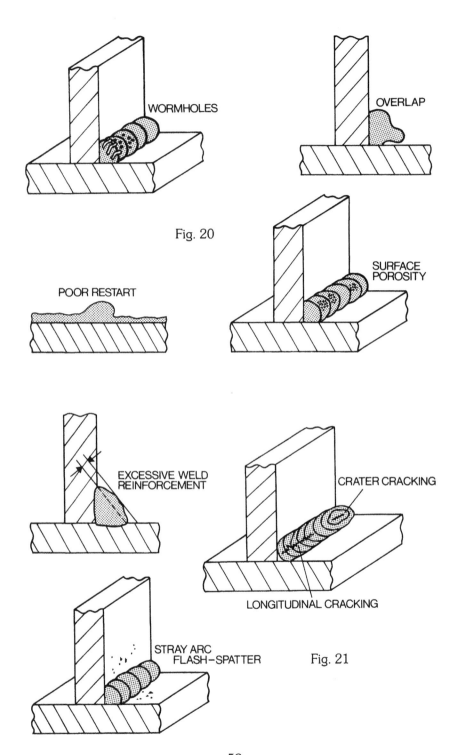

WORMHOLES

OVERLAP

Fig. 20

POOR RESTART

SURFACE POROSITY

EXCESSIVE WELD REINFORCEMENT

CRATER CRACKING

LONGITUDINAL CRACKING

STRAY ARC FLASH–SPATTER

Fig. 21

Magnetic particle inspection

In principle there are two methods of magnetising a specimen:-

i)   The current flow technique.

ii)   The magnetic flow technique.

Both induce a magnetic field into the weld and create magnetic lines of force (flux). Any defect at or near the surface causes flux leakage. Iron powders (inks) are sprayed onto the surface and are attracted to the magnetic 'poles' where the field leaves and enters the steel, at this point the iron oxide particles congregate and this allows us to see the area of particle migration and quite a distinct darker area appears (see Fig. 22).

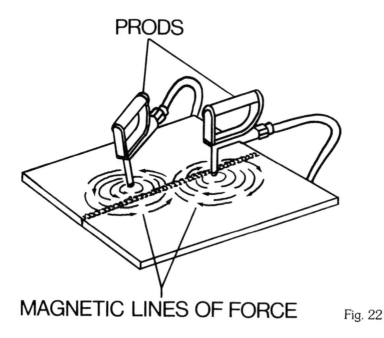

PRODS

MAGNETIC LINES OF FORCE   Fig. 22

Ultrasonic inspection

The ultrasonic beam which is introduced into the plate reflects from any surface in its narrow path. The reflections are shown on a screen with height and spacing according to the reflecting efficiency of the defect and its position along the beam path. The reflecting efficiency will depend on the size and orientation of the defect. A skilled technician can thus interpret the position, and even the size and type of the defect, and decide whether the defect has been caused by a weld defect or whether the ultrasonic beam is simply reflecting from a joint surface.

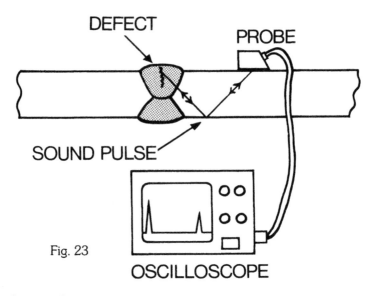

DEFECT

PROBE

SOUND PULSE

Fig. 23

OSCILLOSCOPE

Radiographic inspection

Radiography depends on the image left on film, taken by X or gamma rays which are able to penetrate the weld. As illustrated, a greater intensity of radiation penetrates at locations of flaws. Unfortunately, fine cracking does not effectively reduce the thickness as "seen" by radiography unless the crack is aligned with the X-ray beam. Thus radiography is less effective in detecting fine cracks.

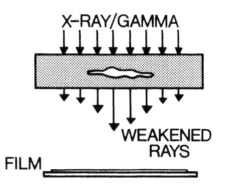

X–RAY/GAMMA

WEAKENED RAYS

FILM

Fig. 24

Macro examination

A cross section is cut from the test weld which is then polished and etched and will show the profile of weld penetration and the heat affected zone. Defects suspected, or detected by other methods can be confirmed and identified. Often the macrosection provides clues on the cause of defects. For example, slag inclusions can be related to particular run shapes or placements, and the nature of cracking can be clarified.

Preparation of surface

The face to be polished should be sawcut then filed with the aim of producing a flat surface with only fine file marks. Remove all sharp edges and stamp any identification required.

Polish the flat filed surface with successively finer grades of wet and dry paper, eg 280, 320, 400, 500. The polishing motion is at right angles to the scratch marks by the previous abrasive and is continued until these disappear.

Etching procedure

The polished surface is degreased using alcohol. Using tongs or rubber gloves the etching solution, which is a percentage of nitric acid in alcohol, is then swabbed over the surface until the weld macro-structure is clearly visible. Without delay the residual acid is washed off with water and the sample is dried. The surface of the specimen is preserved with a light coating of clear lacquer. Any contact between the unprotected etched surface and fingers will cause rapid deterioration.

Break test

If a butt or fillet weld can be broken open, any large defects are likely to be included in the path of fracture. Test welds can be broken to check on weld quality.

Fillet weld fracture test

A T joint with a single fillet is cut to convenient size for breaking under a press or hammer blow as shown in Fig. 25. Fracture will occur through the root to reveal the degree of root penetration, plus any major defects. To encourage fracture through the weld instead of the plate it may be necessary to make a shallow saw cut along the centre of the weld face.

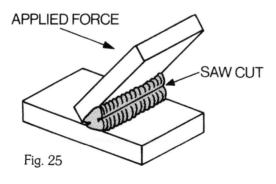

APPLIED FORCE

SAW CUT

Fig. 25

Bend test

The test piece is bent around a former of specified diameter. The former may bend the test piece between two roller supports (see Fig. 26), or a planetary roll may wrap the test piece around the former. The test piece is machined to remove the weld reinforcement and to round the corners slightly.

The weld is usually across the bend test piece so that the outside face during bending is made to undergo a large tensile strain. Even very small joint defects will cause failure so this is essentially a test of joint soundness. However, note that only defects found close to the tension surface of the bend test are revealed. Bend test specimens are selected from the joint so that both sides are tested in separate tests. Thick joints may be tested by a side bend test.

Observations can also be made after the test on weld metal ductility and the matching of weld and plate strength.

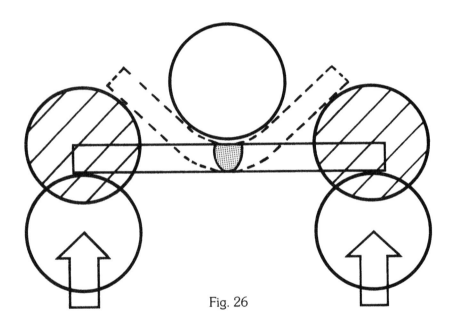

Fig. 26

Hardness test

The hardness of weld zones can be measured by special machines to supply information on the following:-

a) Risk of hydrogen assisted cracking in hardened heat affected zones.

b) Existence of unwanted hard spots due to segregation of alloying elements.

c) Incorrect selection of filler or base metal leading to excessively hard weld zones.

d) Comparison of weld metal hardness with the hardness of the parent metal.

e) Hardness of hard facing required to resist wear on tool or machine surfaces undergoing impact, erosion, high stresses etc.

The hardness testing machine provides a measurement of hardness by measuring the impression made by a standard indentor forced into the metal surface under a standard force. The weld is often sectioned so that the deposit and heat-affected zone can be checked for hard sub-surface areas. The Vickers hardness machine which uses a diamond pyramid indentor is usually used for doing hardness "traverses" on macro-sections. The diamond makes only a small impression so that peak hardness in small areas can be measured. A measurement of the size of the indentation is converted (according to the load on the indentor) to the Vickers hardness number. Other methods of hardness test include the Rockwell (more frequently used for surface tests) and the Brinel hardness test.

Tensile test

These tests are carried out on sample test coupons cut from welds. The strength (ultimate tensile strength) of the joint is tested in a laboratory tensile testing machine using the test piece shown in Fig. 27.

The short length of sample weld is transverse to the direction of pulling. Tension is increased until fracture occurs. Because weld metal strength generally over matches the parent metal, fracture usually occurs in the parent metal.

If the yield strength and ultimate strength of the weld metal are to be measured directly, it is necessary to machine a small tensile test specimen consisting entirely of weld metal from the length of the weld.

Fig. 27

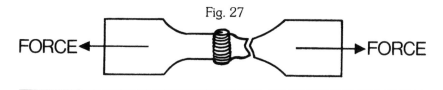

FORCE◄ ►FORCE

CHAPTER 11

PREPARING TO WELD

Aims of this section

The student should understand and identify:-

the importance of, and factors involved in, preparing to weld

the techniques involved in welding joints of poor fit-up

the importance of cleaning

Introduction

The wet-welding process depends very much on the skill of the diver/welder, so before concentrating on any positional welding techniques, there are some basics to be learnt, namely:-

Joint fit-up
Joint preparation
Welding current selection
Electrode angle
Arc length
Travel speed
Recording and monitoring

Joint fit-up

Positioning and fitting must be done with thoroughness and care to ensure a satisfactory weld, remember, if proper care is not taken before welding starts, it will not be possible to achieve a good quality weld.

In fillet welding, it is important that there be no gap at the root of the fillet before welding. If the gap cannot be eliminated, it should be as small as possible.

When a patch or plate is to be used this should marry up as well as possible, ie a concave or convex shaped repair area:- the patch/plate should be shaped to fit.

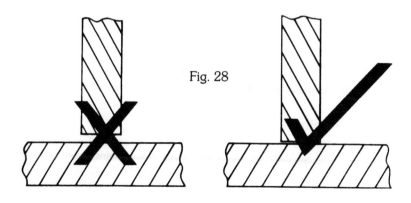

Fig. 28

Joint preparation

The great importance of accurate edge preparation must be appreciated by the student. This is particularly the case where critical control of penetration is required.

Edges must be clean and free of rust and loose mill scale. Edges should be straight square to give no gap. Gaps and rounded edges will cause poor penetration and slag inclusion in the root area.

Welding current selection

The selection of the current should be as described in current ranges under the equipment section.

The student should select the current before entering the dive tank, to give him the nearest correct current value, this will save time as only a fine adjustment will then be needed.

Electrode angle

The electrode angle should generally be between 30 and 45°, although this is not a hard and fast rule. The student should be encouraged to experiment so he gains a greater understanding of what the effects are of changing the angle of the electrode, but, only after he has mastered the technique within the stated angle limits.

Arc length

The welder steadily feeds the electrode into the joint to compensate for burn off. Using the drag technique, the arc length is to a great extent self setting and relates to the current/volts settings and the size of the electrode. With excessive arc length the arc may well extinguish or failing that, the weld metal becomes wider and penetration is reduced.

If the arc is too short the weld bead will become narrow and peaked and the electrode may well stick to the work. To gain experience on the effects of arc length, make runs with too short and too long arc lengths.

Travel speed

Concentrate on maintaining an even travel speed to give the correct width of run. In fillet welds this is measured by the leg length, so a 4mm electrode should give a 4mm leg length approximately. The rate of travel will depend on:-

> the electrode size
> the current
> the run size required.

Surface cleaning

It is most important to clean the surface to be welded properly, as a satisfactory weld cannot be achieved over thick paint, rust or marine growth. Even the initiation of an arc may be impossible, or at best, very difficult.

In multiple pass welds, each bead must be thoroughly cleaned before depositing the next one. Patches or plate should be prepared/cleaned above water.

The use of pneumatic tools are of great help underwater to assist in cleaning, chipping, peening and grinding.

Recording and monitoring

This is an important job, both the diver and his tender need to work together. Each man needs to know what is expected of him and what information needs to be recorded. During the course of the dive, the diver and tender will be giving instructions and asking questions. This relevant information needs to be recorded, so that a procedure can be compiled. This will highlight problem areas and hopefully provide answers as to what needs to be changed next time. It will also provide information on what helped to make a good weld.

Here is a list of some typical instructions and questions:

1. Have the electrode made hot and cold
2. Have electrodes sent down, usually in small bundles of 5-10.
3. Constant amperage readings.
4. Give information on the state of the prep and the condition of the general area. Even possibly video this information.
5. Travel speeds (ROL) timings.
6. Have the current adjusted.
7. Give information as to how many passes have been completed (because it is very easy to forget).
8. Give information as to whereabouts on the job he is.
9. Whether some runs are much better than others.
10. How much has been completed.
11. Whether the earth clamp(s) have had to be moved around the job.
12. Any magnetic problems, when and where they occurred.
13. Any difficulties.

Poor fit-up

When poor fit-up is the case, additional weld metal is needed to fill any gaps that may be present at the root of the weld. This can be done by a technique called "feeding in".

This is accomplished by feeding the electrode into its own arc/molten pool, together with a forward and backward movement. It must be emphasised that, as stated earlier within this manual, good fit-up is essential and this technique is only to be used if the fit-up cannot in any way be improved, as wet-welding techniques in poor fit-up situations are not entirely successful. However, having said that, this technique may be successful if the gap is no more than 5-8mm (see Fig. 29).

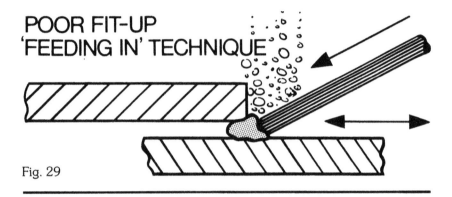

POOR FIT-UP
'FEEDING IN' TECHNIQUE

Fig. 29

CHAPTER 12

PRACTICAL WET-WELDING EXERCISES

During this chapter you will be reading about the practical techniques that Hydromech Technical Services has programmed together, that will enable you to put into practice all that you have read in this manual, at our own in-house facilities.

Index to exercises
1.   Striking the arc/breaking the arc.
2.   Electrode manipulation.
3.   Pad weld – flat position.
4.   Outside corner fillet weld flat position – single pass.
5.   Fillet weld – flat position – single pass.
6.   Fillet weld – horizontal-vertical – single pass.
7.   Fillet weld – pipe to plate horizontal-vertical – single pass.
8.   Fillet weld – horizontal-vertical – multi-pass.
9.   Fillet weld – vertical-down – single pass.
10. Fillet weld – vertical-down – multi-pass.
11. Fillet weld – overhead – single pass.
12. Fillet weld – overhead – multi-pass.
13. Fillet weld – pipe to plate horizontal-vertical – multi-pass
14. Poor weld fit-up technique

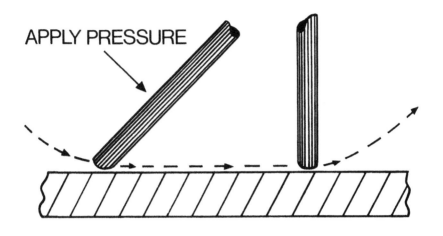

APPLY PRESSURE

Fig. 30

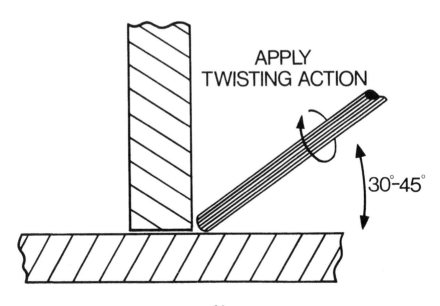

APPLY
TWISTING ACTION

30°-45°

Exercise 1: Striking the arc/breaking the arc

Clamp the electrode in the holder. Having made sure the electrode is COLD, position the earth clamp so any welding is travelling towards the earth.

Make yourself as comfortable as possible. Do not hold the electrode holder too tight, or your arm too rigidly. Relaxed control will not be as tiring and will help you to hold the electrode steady.

Arrange the welding lead so that the load is taken off your hand and wrist. Either loop the cable over your shoulder or tie off the end to a suitable object. DO NOT LOOP THE CABLE AROUND YOUR BODY, as in the event of an emergency you may become entangled.

Now call to make the electrode HOT. Place the electrode on the work piece and with your other hand apply pressure onto the electrode and now with the hand holding the electrode holder draw away as if striking a match. This should have removed any flux from the end of the electrode. It may or may not have arced, in either case you are ready to start your first pass. Prepare yourself to start your weld because this time when the electrode touches the work it will start to be consumed.

Look very carefully where you are to start, place the electrode at that point, make contact. At the same time, apply a small twisting action making sure you have the electrode at the correct angle (see Fig. 30).

At the end of the weld run withdraw the electrode from the material quite quickly. Alternatively, slow down your forward travel when nearing the end of the weld run, stop for a short pause to help fill any crater which may have formed, then remove the electrode in a J sweeping fashion, similar to the striking of the arc technique.

Note: It will soon be realised by the students that on many occasions they will not be able to see when they are nearing the end of the weld run. So a close look at the joint before any welding starts to get an impression of the approximate length required will help.

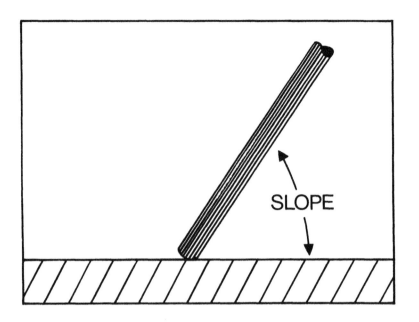

Fig. 31

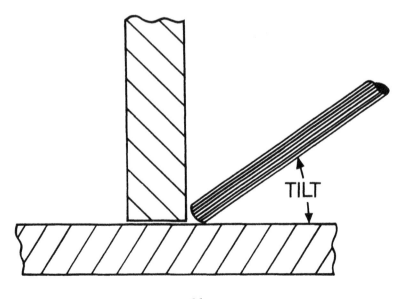

Exercise 2: Electrode manipulation

The electrode shall be simply travelled in a straight line. The angle of the electrode is controlled by the welder so that the arc direction assists in the control of the weld pool.

The electrode angle is measured between the electrode and the direction of travel so that angles less than 90° indicate a trailing technique being used, and angles greater than 90° indicate that the tip of the electrode is pointed fowards. The electrode is usually run at a trailing angle so that the force is directed as near as possible to the base metal at an angle which best facilitates metal transfer.

Run placement

In multi-pass welding runs, weld beads or stringers are placed so that succeeding runs have good access to fusion faces. Run placement should be designed so that a shelf for the following run is made.

a) Position the electrode at an angle of approximately 30° to the line of the weld with the electrode tip in contact with the work.

b) Call for making the electrode HOT, scrape the tip of the electrode to remove the waterproof coating to start the arc.

c) Apply sufficient pressure in the direction of the arrow to allow the electrode to consume itself (see Fig. 31).

Note: The angle of the electrode should generally be between 30 and 45°.

Exercise 3: Pad weld – flat position

Aim: To produce a pad weld with a smooth surface. The welding shall extend to the edge of the plate.

Objectives: To show the student the difficulties in running weld beads and keeping the welds parallel without the help of any natural grooves.

Preparation: *Material* – 10mm steel plate
1 off 100 x 100mm

Electrode and amperage:

Electrode size, mm	Amperage range
3.25	145-180
4.00	170-210

Procedure: To deposit the first pass down the centre of the plate and subsequent passes with overlap until the end of the plate is reached, in both directions.

The student should keep a mental note of exercises 1 and 2 during this exercise.

After each weld bead, chip slag off and wire brush until weld metal surface is clean before running next pass.

Evaluation: Inspect visually for surface defects, such as undercut, poor profile, porosity, cracks, etc, and for any electrode wandering problems.

Exercise 4: Outside corner fillet weld – flat position – single pass

Aim:	To produce a corner fillet weld in the flat position.
Objectives:	To show the student the difference when welding a preparation that has a natural groove for the electrode to follow.
Preparation:	*Material –* 5mm steel plate 2 off 50 x 200mm

Electrode and amperage:

Electrode size, mm	Amperage range
3.25	145-180
4.00	170-210

Procedure: Firstly, follow exercises 1 and 2.

When the arc has started exert enough pressure against the work to allow the electrode to consume itself. In this instance your travel speed should be slower than in the previous exercises, because you are practising building up the weld metal to fill the prep. This prep is ideal in helping you achieve a good weld because it affords you a natural groove. However, still make mental notes on electrode angles and travel speeds. Compare the difference in your mind between this type of prep and the prep in other exercises.

Make notes and compare how much easier or harder this prep seems to you. You should not have any electrode wandering problems with this exercise, but if you do, look back at all the events prior to starting to weld and during welding, you may find where you went wrong.

Evaluation: Check the size of the weld. Did you fill the prep up in one pass or not?

Inspect visually for surface defects, such as undercut, porosity, cracks, etc. Did you over fill the prep? If so, your travel speed was too slow.

Exercise 5: Fillet weld – flat position – single pass

Aim: To produce a 4 and 6mm fillet weld in the flat position ensuring that the weld has equal and correct leg length with no undercut. Contour should be smooth with good blending at the toes of the weld.

Objectives: To show the student how to make a single pass mitre fillet weld.

Preparation: *Material –* 10mm steel plate
2 off 100 x 200mm

Assemble plates to make a T joint. All tacking to be done in air. Avoid any gaps at the root. Support the plates for welding in the correct position.

Electrode and amperage:

Electrode size, mm	*Amperage range*
3.25	145-180
4.00	170-210

Procedure: Firstly, follow exercises 1 and 2.

When the arc has started exert enough pressure against the work to allow the electrode to consume itself. Maintain a constant travel rate and ensure the electrode remains in contact with the work piece at all times. Always make mental notes of the electrode angle(s) and try to keep in the range 30-45°. Concentrate on trying to obtain a mitre fillet weld. Don't rush the electrode, try and allow it to set the pace. Run straight beads, do not weave. When finished call to make it cold. Chip slag off and wire brush weld ready to inspect.

Evaluation: Check the size of your fillet.

Inspect visually for surface defects, such as undercut, poor profile, porosity, cracks, etc.

Saw out a sample approximately 100mm long to carry out a fillet weld fracture test. Examine the fracture surface for any defects, check for root penetration.

Exercise 6: Fillet weld – horizontal-vertical – single pass

Aim: To produce a 4 and 6mm fillet weld in the horizontal-vertical position, having equal and correct leg lengths, with no undercut, porosity, or unfilled craters. The weld should have good profile. On this exercise the student should make a stop/start in the centre of the plate. This stop/start should blend in with the weld contour.

Objectives: To show the student how to make a mitre fillet weld in horizontal-vertical position, making sure that the weld's leg lengths are equal.

Preparation: *Material –* 10mm steel plate
2 off 100 x 200mm

Assemble plates to make a T joint. All tacking to be done in air. Avoid any gaps at the root. Support the plates for welding in the correct position.

Electrode and amperage:

Electrode size, mm	Amperage range
3.25	145-180
4.00	170-210

Procedure: Firstly, follow exercises 1 and 2.

When the arc has started exert enough pressure against the work to allow the electrode to consume itself. Maintain an even travel speed. When trying to weld in this position you are actually trying to deposit half of the weld metal onto a vertical face, so correct electrode angle is vital. Try to maintain a mitre fillet. Let the electrode set the pace. Make straight runs, do not try to weave. When finished call to make it cold. Chip off slag and wire brush weld ready to inspect.

Evaluation: Check the size of your fillet. Is it a mitre fillet? Does it have equal leg lengths? Are they the correct size?

Visually inspect the surface.

Carry out fillet weld fracture test, examine fracture surface for any defects, check for root penetration.

Exercise 7: Fillet weld – pipe to plate – horizontal-vertical – single pass

Aim:
To produce a 4 and 6mm fillet weld in the horizontal-vertical position joining a pipe to plate.

Objectives:
To give the student practice in trying to maintain correct electrode angles and travel speed, etc, when faced with a contoured component.

Preparation:
Material –   1 off 10 x 150mm OD pipe
1 off 250 x 250mm plate
Support in correct position

Electrode and amperage:

Electrode size, mm	Amperage range
3.25	145-180
4.00	170-210

Procedure:
Firstly, follow exercises 1 and 2.

Try and break the pipe up into four quarters, and concentrate on achieving a good weld in each quarter. Stop at each quarter if you are having problems – it is all good stop/start practice. Otherwise try to complete as much as you can. Welding pipe to plate is virtually the same procedure as the other exercises. But this time the electrode has got to move around the pipe, so this means you are going to have to move as well. So, get comfortable and concentrate on maintaining the correct electrode angle, and travel speed, as well as producing a mitre fillet weld.

You may find this exercise the hardest yet, so consider all that you have learnt and simply implement it.

When finished make it cold. Chip off slag, wire brush ready to inspect.

Evaluation:
Check the size of the weld and your stop/start positions. Is the weld a mitre fillet? Inspect visually for any defects. Make your own evaluation. Are you satisfied? If not, why not – try again.

Exercise 8: Fillet weld – horizontal-vertical – multi-pass

Aim: To produce a 6mm three run pass fillet weld, with good weld profile and weld overlap.

Objectives: To give the student practice in making multi-pass welding runs, having equal leg lengths.

Preparation: *Material –*   10mm steel plate
2 off 100 x 200mm

Electrode and amperage:

Electrode size, mm	*Amperage range*
3.25	145-180
4.00	170-210

Procedure: Firstly, follow exercises 1 and 2.

Having deposited your first pass, chip off all the slag and wire brush the weld. Make absolutely certain that you have removed the spatter and blended in any high or low areas. Then run your second pass on the bottom, first making sure you overlap the first pass by 50%. Then your third pass will have a shelf to sit on which will help you in guiding the electrode on your final pass. When finished make it cold, chip off slag, wire brush ready to inspect.

Evaluation: You will probably have noticed that it is very difficult to make your second and third pass if your root run was of poor profile. And the importance of cleaning and blending is vital, not only to improve your performance, but the welds too.

Exercise 9: Fillet weld – vertical down – single pass

Aim: To produce a mitre fillet weld when welding vertical-down.

Objectives: To show the student the difference when welding in the vertical position.

Preparation: *Material –*   10mm steel plate
2 off 100 x 200mm
Support plates in correct position

Electrode and amperage:

Electrode size, mm	Amperage range
3.25	145-180

Procedure: Firstly, follow exercises 1 and 2.

When welding vertically with wet-stick, we always weld in a downwards direction. This is so the bubbles created by the arc do not interfere with your visibility. You may have to manipulate the electrode more to assist you in making your pass.

Before actually striking the arc, make sure that there are no obstructions in your way to stop you completing the first pass in one go, ie make sure you can maintain the correct angle of the electrode all the way to the bottom, remembering your hand/arm and the electrode holders lead might get snagged. Once you have started, do not stop, maintain your travel speed. You may find that you have to travel that bit faster when welding vertically down. Travel speed is very important when welding vertically down. When finished make it cold. Chip off slag, wire brush ready to inspect.

Evaluation: Check whether you maintained root fusion on both faces. Is the weld profile poor? If so, did you have difficulties in maintaining your travel speed? Did the weld try to overtake you? Did you have the correct electrode angle? If you are not happy with the weld perhaps you need to check all the procedures carefully again. Then try again.

Exercise 10: Fillet weld – vertical down – multi-pass

Aim: To produce a 6mm three run pass fillet weld, with good weld profile and weld overlap.

Objectives: To give the student practice in making multi-pass vertical-down welds.

Preparation: *Material* –   10mm steel plate
2 off 100 x 200mm
Support plates in correct position

Electrode and amperage:

Electrode size, mm	Amperage range
3.25	145-180
4.00	170-210

Procedure: Firstly, follow exercises 1 and 2.

Having deposited your first pass, chip off all the slag and wire brush the weld. Make absolutely certain that you have removed the spatter and blended in any high or low areas. Then run your second pass on the left side making sure you overlap the first pass by 50%. Then run your third and final pass on the right side, making sure that the toes of the second and third pass meet in the central area of the first root pass. When finished make it cold, chip off slag, wire brush ready to inspect.

Evaluation: As Exercise 7/8

Exercise 11: Fillet weld – overhead – single pass

Aim: To produce a 4mm fillet weld in the overhead position.

Objectives: To show the student the difficulties in producing a weld of good profile in the overhead position.

Preparation: *Material –*   10mm steel plate
2 off 100 x 200mm
Support plates in correct position

Electrode and amperage:

Electrode size, mm	Amperage range
3.25	140-170

Procedure: Firstly, follow exercises 1 and 2.

In overhead wet-stick welding the current setting is much more critical than in other positions, the range of the current is that bit narrower. So, it is vital that the student selects, as near as possible, the correct current before entering the tank.

Position the test plates in the overhead position at an angle of approximately 45°. Then follow procedures as in previous exercises.

Evaluation: This is regarded as one of the most difficult positions and it will test whether the student has fully understood all he has learnt in previous exercises to the full. Inspect welds for surface defects.

Exercise 12: Fillet weld – overhead – multi-pass

Aim: To produce a 6mm fillet weld in the overhead position.

Objectives: To give the student practice in making multi-pass welds in the overhead position.

Preparation: *Material –* 10mm steel plate
2 off 100 x 200mm
Support plates in correct position

Electrode and amperage:

Electrode size, mm	Amperage range
3.25	140-170
4.00	170-190

Procedure: Firstly, follow exercises 1 and 2.

Having deposited your first pass, chip off all the slag and wire brush the weld. Make absolutely certain that you have removed the spatter and blended in any high or low areas. Then run your second pass on the bottom, first making sure you overlap the first pass by 50%. Then your third pass on the top of the root pass, again making sure that 50% has overlapped. When finished make it cold, chip off slag, wire brush ready to inspect.

Evaluation: Check whether correct leg lengths and weld profile are in order. Inspect visually for any defects, and make your own evaluation.

Exercise 13: Fillet weld – pipe to plate – horizontal-vertical – multi-pass

Aim: To produce a 4 and 6mm fillet weld in the horizontal-vertical position joining a pipe to plate.

Objectives: To give the student practice in making multi-pass welding runs when faced with a contoured component.

Preparation: *Material –*   1 off 10 x 150mm OD pipe
                                  1 off 250 x 250mm plate
                   Support plates in correct position

Electrode and amperage:

Electrode size, mm	Amperage range
3.25	145-180
4.00	170-210

Procedure: Firstly, follow exercises 1 and 2.

                   Follow a multi-pass exercise.

Evaluation: Are you satisfied? If not, why not – try again.

Exercise 14: Poor weld fit-up technique

Aim: To produce a lap weld in the horizontal-vertical position.

Objectives: To show the student the technique of welding when faced with a poor fit-up.

Preparation: *Material* – 10mm steel plate
2 off 100 x 200mm plate
Assemble plates together by lapping by 20mm.

Electrode and amperage:

Electrode size, mm	Amperage range
4.00	170-210

Procedure: Firstly, follow exercise 1.

When poor fit-up is the case, additional weld metal is needed to fill the gap at the root of the fillet. This can be done by "feeding in" the electrode towards the joint faster than usual, together with a back step technique. When using this technique no travel speed procedure is used. The object is to deposit weld metal to fill the gap by applying pressure in a forwards direction while the electrode is being consumed.

Evaluation: This is a very difficult procedure and can only achieve limited success, and this depends upon the size of the gap which has to be filled. One should not expect a weld with good profiles.

Printed and bound by CPI Group (UK) Ltd, Croydon, CR0 4YY

03/10/2024

01040437-0002